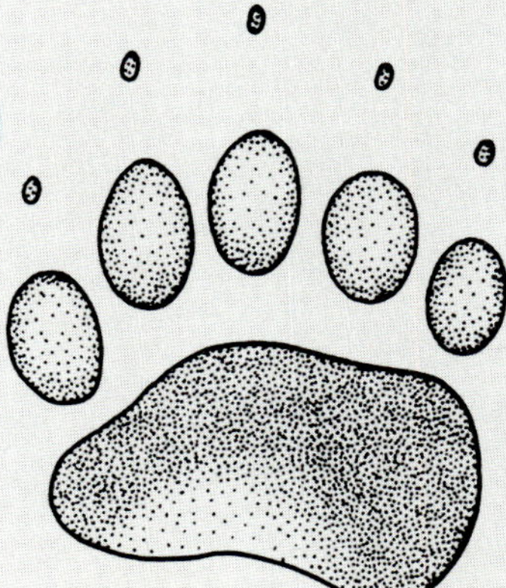

Brown bear family, McNeil River

Caribou

Ranger tarandus

Range/Habitat

Caribou live on the arctic tundra near or above timberline, particularly places with rolling foothills and mountainous areas. There are four varieties of caribou, but only the barren ground caribou is found in Alaska (they also range across Canada east to Hudson's Bay). In general, barren ground caribou are not found in the far Southeast panhandle or on offshore islands of the Bering Sea (but some have been transplanted to islands such as Adak). There are approximately 800,000 caribou in Alaska living in about 35 different herds.

Physical description

Adult males height at the shoulders varies from 3.5 to 5 feet. Males can weigh up to 700 pounds but 275 to 400 pounds is average; females are smaller, averaging 175 to 225 pounds.

Color varies but the most common pattern is brown with white on the neck, under the tail, on mane and belly.

Caribou are very unusual in that both males and females have antlers. Male antlers can reach up to 58 inches wide and most are dropped in the fall after the rut. Females keep antlers all winter and lose them when they calve in the spring. Non-pregnant females and young drop antlers in April to May.

Caribou make a peculiar clicking sound when they walk; this is due to the movement of tendons and small bones in their legs.

Life cycle

Caribou are members of the deer family and graze on willows, birch, fungi, grass and other plants in summer. In fall when the plants die they switch to mosses and lichens. Rezendes reports some caribou have been seen feeding on dropped antlers and lemmings.

Caribou are very gregarious and herds can vary from 500 to more than 300,000 animals. Such large groups of grazing animals would soon deplete their food supply; fortunately caribou migrate across hundreds of miles and use is distributed. In the summer, the males and young will further disperse into smaller groups.

September is rutting season during which bulls use up their summer fat stores in the competition for females (and guarding and competing for them day and night); they have little time to put weight back on for the winter.

In early spring the herd will begin its migration to calving areas and summer range. In one of the most famous of arctic treks, the Porcupine herd migrates 800 miles from the southern Yukon north to Arctic Ocean.

A single calf is born in late May or early June and can walk within an hour. Browne reported observing most females in a herd in Denali Park dropping their calves within a few days of each other.

In Alaska, grizzly bears will hunt calves systematically at calving time. Caribou depend on their sight to detect danger and can distinguish between hunting and non-hunting wolves. Their primary means of escape is running; a caribou can gallop up to 40 mph.

Northern survival

Hair formed between the toes provides extra traction across ice; hoofs spread apart to support weight in snow and on soft tundra. Extra hair grows on muzzles for winter. Long, hollow, air-filled guard hairs cut the wind and help provide warmth.

Tracks

Tracks are crescents 3 to 4 3/4 inches long by 4 1/2 to 5 3/4 inches wide. Caribou have dewclaws, a soft inner part of the hoof that hardens or shrinks; these apparently provide extra support on the snow.

Other signs

Antler rubs, scat (pellets similar to deer and elk)

Caribou trivia

The name caribou means "pawer" or "scratcher" and refers to the holes made in the snow as they forage for lichen.

Caribou may be confused with reindeer (domesticated Eurasian caribou); don't worry as they are actually the same species.

Caribou are very intelligent in trying avoid insects; they will move to snowbanks, ridges, and the Arctic coast to catch breezes.

Where to see
Denali National Park
Arctic National Wildlife Refuge
Lake Clark National Park
Denali, Dalton and Glenn Highways

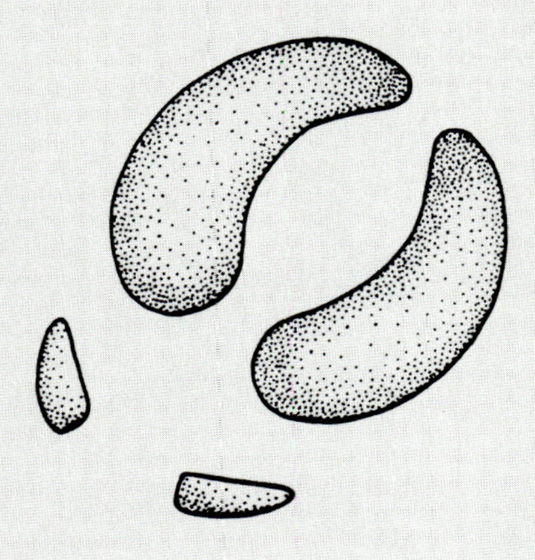

Caribou

Dall sheep

Ovis dalli

Range/Habitat

Dall sheep are found in rugged alpine areas. Here they find food and raise their young along isolated ridges and meadows. The Alaska population numbers about 70,000 (1989); half of these are in the Brooks Range.

Physical description

Adult males average 5 feet in length and weigh up to 250 pounds; females are slightly smaller, averaging 4 1/2 feet long and weighing 150 pounds.

Both sexes have horns and it is hard to tell them apart until they are three years old when the males get larger. Horns stop growing in winter and therefore have rings (called annuli) which biologists use to easily determine their age. A sheep 12 years in age is considered old in the wild; they have lived up to 19 years in captivity.

Life cycle

Lambing takes place in special areas on southern exposures of steep cliffs to avoid their major predators: wolves, bears, wolverines and golden eagles. Lambs are born in late May or early June; a single lamb is the usual case. Sheep depend on their sense of sight to keep them out of danger.

Sheep will move between summer and winter ranges, but these are not far apart. They feed on grasses, sedges and forbes in the summer but have to switch to frozen grass, moss and lichen in winter.

Sheep have a very well developed social system stay together in groups. Social order is determined by the famous head bashing contest where two same-sized males meet each other. Rezendez reports they can ram one another from distances of 30 feet and reaching speeds exceeding 20 mph. Fortunately they have layered skulls that absorb the shocks.

Northern survival

Sheep are excellent jumpers and are very agile and surefooted on their home ledges and talus slopes where they can easily escape predators. Some have been observed jumping off a ledge 20 feet high. In winter, winds blow snow from forage.

Tracks

Cloven hoof pads have rough bottoms for extra traction. The tracks have rounded tips and straight, rectangular sides. Male tracks average 3 1/2 inches long by 2 1/2 inches wide. Ewes' are about 1 3/4 to 2 1/8 inches long by 1 1/2 to 1 3/4 inches side. The rear feet will hit slightly off to the side of the front feet, leaving a slightly double-track appearance to their trail.

Other sign

Look for bedding areas (a scraped, oval depressions) along cliffs or in caves.

Sheep trivia

Sheep will travel long distances to reach mineral licks; such licks are often used by several different bands and intermingling can occur. This increases the genetic diversity of each band.

Where to see

About half the Dall sheep in the state inhabit the remote Brooks Range. Other more accessible locations to find them include the Alaska Range (such as in Denali National Park), the Wrangell Mountains, Chugach Range and the mountains on the Kenai Peninsula. One popular place is McHugh Creek along the Seward Highway (mile 106.5) near Anchorage.

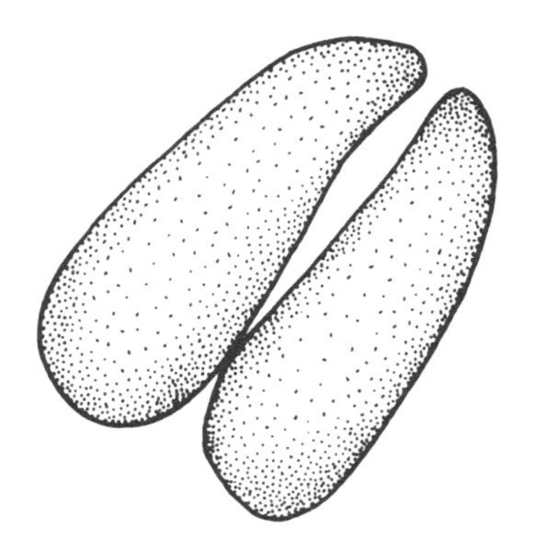

Dall sheep

Red fox

Vulpes vulpes

Range/Habitat
Red foxes are found in most of Alaska except some of the southeast islands, the western Aleutians and Prince William Sound. However, fox farming operations in the early 1900s introduced them to many remote islands, particularly in Prince William Sound. Red foxes like lowlands and hills. While they will share ranges with arctic fox, where both occur the red fox seems to dominate; there are reports of them digging arctic fox from their dens and killing them.

Physical description
Adult red foxes average about 22 to 32 inches long in the body plus a white-tipped tail 14 to 16 inches in length. Weight averages six to 15 pounds; males are heavier than females. Their coat is light yellow to red with black markings on the bottom half of the legs. Red foxes have a variety of color phases and can have varying amounts of black and silver in their coat.

Life cycle
Foxes breed in February and March. Young are born 53 days later in a den on the side of a small hill dug. An average sized litter is four, but it can be as many as 10. Young learn to hunt in three months; the family breaks up in autumn.

Red foxes have well developed sight, smell and hearing that enable them to catch and eat small mammals (particularly voles), birds, insects, eggs, and carrion. Enemies include wolves, man, lynx and wolverines; kits have been attacked by eagles.

Northern survival
Red foxes will cache (bury) extra food when it's abundant.

Tracks
Tracks are 2 1/8 to 2 7/8 inches long by 1 5/8 to 2 1/8 inches wide (front). There is a distinctive line in the heel of the front track (usually straight across but it can also be boomerang-shaped).

Other sign
Dens, scent posts, scat

Fox trivia
Foxes are classified with dogs but have eyes similar to cats (dilate elliptically up and down).

Where to see
Red foxes are generally secretive and cautious but those in Denali National Park have grown accustomed to people and are frequently observed from shuttle busses along the main park road.

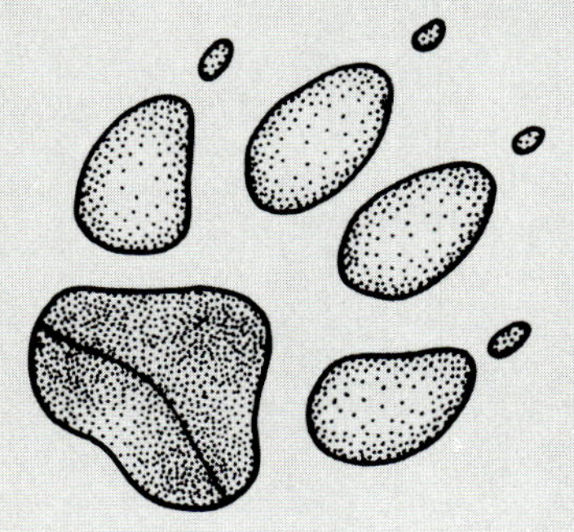

Red fox mother and kit

Arctic fox

Alopex lagopus

Range/Habitat
Arctic foxes are found in the treeless coastal areas in western and northern Alaska, including parts of the Alaska Peninsula, Aleutian Islands and islands of the Bering Sea.

Physical description
Adults weigh from six to 10 pounds. Size averages 30 inches in length plus an additional 15 inches for the tail. Arctic foxes are white in winter changing to brown on the back in spring. A blue phase is common in the Aleutian and Pribilof islands; these remain dark all year but will lighten in winter.

Populations fluctuate greatly but generally these animals are relatively scarce as about 4,000 are estimated to exist statewide (1989).

Life cycle
Foxes mate in March and April. Fox kits are born 52 days later in a den built in low mounds or river banks. Dens usually have a southern exposure and may be 12 feet deep; multiple branches and exits are common. An average litter is seven but as many as 15 can occur. Both parents provide food to the newborn but the young can hunt with them in three months. The family breaks down by fall after which each member lives on its own.

Fox are adaptable in their diet, seeking carrion, berries, and small mammals such as lemmings, ground squirrels, and rodents. In coastal areas, they will seek birds and bird eggs. Enemies include the gray owl and snowy owl; bears and wolves have been known to dig foxes out of their dens and kill them.

Northern survival
Short legs and short ears conserve heat. White winter color lets it get close to its prey.

Tracks
Tracks are 2 1/8 to 2 7/8 inches long by 1 5/8 to 2 1/8 inches wide (front).

Fox trivia
Many arctic foxes venture out on the pack ice to scavenge on seals killed by polar bears. They also will kill gulls and auklets that come ashore to rest.

Where to see
Coastal areas around Nome, Unalakleet, and Barrow.

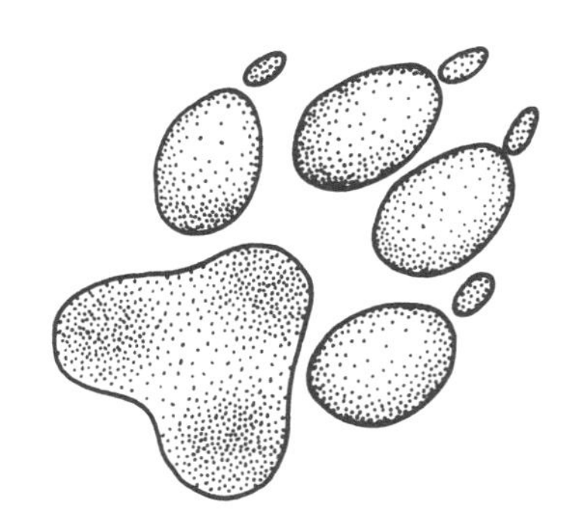

Arctic fox

Major land habitats in Alaska

You can improve your chances of seeing wildlife by learning to recognize the major natural habitats in Alaska and knowing the habitat preferences and habits of the animals that live there.

Tundra

Tundra has been described by one writer as "a land as vast and lonely as the sea." This likely refers to the millions of acres of treeless plains found north of the Brooks Range and along the west coast of Alaska. Permafrost keeps water from draining, leading to the formation of numerous lakes, ponds and soggy ground which are called *wet* or *moist tundra.* It is home for grizzly bears, musk ox, foxes and millions of birds who come to nest and raise their young. The drier *alpine tundra* commonly appears as meadows on ridgetops above timberline. Vegetation consists mainly of slow-growing, low plants interspersed with rocks. Low-growing shrubs such as blueberries, cranberries and willows along with sedges, grasses and lichens provides food for the animals that live or visit there. Look for Dall sheep, caribou, marmots and pikas. Alpine tundra is one of the best places to see wolves, brown bears, wolverines and red foxes because of the wide open spaces and vistas.

Spruce-hardwood forest

Spruce-hardwood forest habitats consist of black spruce, white spruce, birch, aspen, poplar and willow in varied combinations throughout interior Alaska. Black spruce are often stunted due to the harsh growing conditions. Lightning-caused wildfires create a mosaic of trees and understory plants that provide a variety of animal habitats. The growth stage of the forest will influence the types of animals you can see. If there are many shrubs and saplings, watch for moose, fox, wolves and bears. Hares will inhabit meadows and in turn attract lynx. Old growth has the fewest species but is important for the cover it provides.

Dominant habitats

1. Wet and alpine tundra
2. Spruce-hardwood forests
3. Hemlock-spruce forests

Evergreen hemlock / spruce rain forest

Warmer temperatures and higher precipitation along the coast allow for the development of the evergreen hemlock and spruce rain forest. This provides a home for deer, bears, flying squirrels, bald eagles, owls, woodpeckers, wrens and various songbirds.

Muskeg

Muskegs occur in depressions where old sloughs and lakes formerly existed. Water concentrates on the surface because muskegs are commonly underlain by permafrost that limits drainage. Plants include sphagnum mosses, sedges, lichens, Labrador tea, willow, cranberry and blueberry.

Scattered black spruce trees occupy drier sites within muskegs. Individual trees frequently lean because the ice below them causes frost heaving and limits the root depth needed for adequate support.

Muskegs are excellent sites for collecting berries, but walking between the narrow channels of melting ice and the raised mounds of vegetation can be as difficult as running an obstacle course.

Moose use the dense shrub thickets to deliver their young and hide them from predators. Also look for beaver, muskrat, weasels and mink. Loons, grebes and other birds may be seen in open water.

Riparian habitat

Riparian habitat consists of riverbank and lakeshore vegetation. It is usually more lush than adjoining inland vegetation because it's closer to water and nutrients. Wildlife depend heavily on riparian habitat because of the variety of food and living space created at the common boundaries of aquatic and terrestrial plant communities.

White spruce, cottonwood trees, tall willows and dense woody shrubs are common. These plants provide food for moose and homes for mice and shrews as well as for the furbearers (such as weasels) that feed on them. Mature trees along the water's edge are used as nesting sites by a variety of birds such as bald eagles, osprey, woodpeckers, hawks and owls. Songbirds may be seen in small trees and shrubs.

Lynx

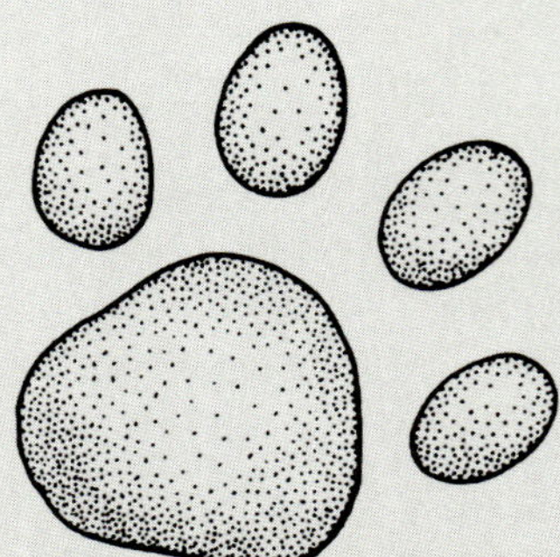

Lynx

Felis lynx

Range/Habitat

Lynx are found throughout Alaska except for the Aleutian Islands, various offshore islands and some islands in Prince William Sound and Southeast Alaska. Look for them in spruce and hardwood forests, and in mixed shrub communities and meadows where plants are young (this is prime habitat for their food of choice, snowshoe hares).

Lynx are shy and unobtrusive but are not necessarily scarce; populations in a given area are strongly influenced by the abundance of snowshoe hare and other small game. If food is difficult to find, fewer females breed and fewer kittens survive winter. Lynx will migrate out of areas where food is scarce.

Physical description

The lynx is a large buff-colored cat with a very short tail. Legs are long and the feet are very furry. Unique distinguishing characteristics are the black-tipped tail and long tufts on the ears. Adults average 18 to 30 pounds, males being the heavier and occasionally reaching 40 pounds.

Life cycle

Lynx mate in March and April. Kittens are born in 63 days; litters average two to four but can reach six depending on population dynamics. Females giving birth will typically seek a sheltered area such as a rock ledge or under a fallen tree. The family breaks up after a year with each yearling seeking its own territory.

Lynx typically have a home territory of up to 100 square miles that they will travel about, covering one to five miles a day in search of prey. They will either hope for a chance encounter or patiently sit and wait along a small game trail. In either even they run down the prey in a burst of speed.

The favorite food of lynx is the snowshoe hare, but lynx will also switch to fox, ptarmigan, grouse and squirrel if hares are in short supply. If food really gets scarce they have been known to take on caribou and Dall sheep. A lynx prefers fresh meat, not carrion, unless it's really hungry.

Northern survival

Lynx will migrate out of food-poor areas. Their light body weight coupled with thick hair on the bottom of their feet makes for hairy paws that keep them on top of the snow.

Tracks

Lynx have five toes on the front feet but one does not register; watch for a double lobe on the leading edge of the heel pad. Claws are drawn up into toe pads when walking or running. Feet are3 1/4 -3 3/4 inches long by 3 - 3 3/8 inches wide for front; hind feet are slightly smaller.

Other sign

Lynx go over the same route, learning to recognize the trail, then staking out a spot to wait. They will hole up under ledges and timber deadfalls to stay out of the weather or have their young; look for scat in these areas. Lynx love ledges on steep hillsides where views are good.

Lynx trivia

"Lynx" is Latin for "lamp" and refers to the large iris in their eyes which enables them to hunt well at night; their eyes also have reflectors to take advantage of all light available.

Where to see

No specific locations can be recommended. Review their general habitat types, habits, tracks and other signs. Some people claim it is possible to attract lynx by appealing to their curiosity by hanging a shiny object along one of their trails. Lynx are rarely active in the day, preferring to hunt in twilight or night.

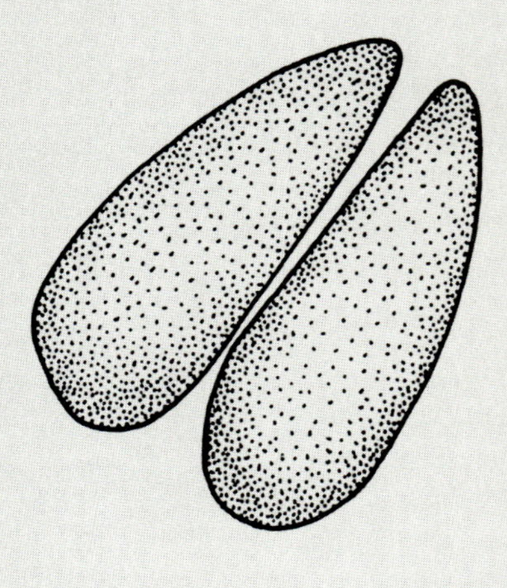

Moose

Physical description

The moose is the largest member of the deer family and Alaska moose (*Alces alces gigas*) are the largest moose in the world. Their color varies from brown to almost black and will change in an individual with the season and the age of animal. Adult males average 1,200 to 1,500 pounds. Adult females generally average 800 to 1,300 pounds. Calves weigh 22-35 pounds at birth and are red-brown to light rust in color.

Only bulls have antlers, which are broad-bladed and can spread to six feet or wider; antlers are shed in December and January.

Moose are ungainly and sometimes said to be the perfect example of "an animal designed by a committee." A ropelike flap of skin (called a dewlap or bell) hanging from neck, long legs, a high, humped body, the downward-turned nose, and big ears help contribute to this impression. But the legs help them travel through deep snow and the nose has flaps of skin that seal out water while it feeds underwater. No one knows what, if anything, a dewlap is for.

Range/Habitat

Moose are distributed in suitable habitat from the Stikine River in Southeast Alaska north to the Colville River on the Arctic Slope. Their distribution extends southwest to port Moller on the Alaska Peninsula. Moose can often be seen along rivers and in burned areas where they can find a good supply of willow and birch for food. They will also feed in shallow ponds.

Moose

Alces alces

Life cycle

Moose breed in fall. Bulls do not assemble harems; instead they will stay with one female at a time but can breed more than one female a season. Calves are born in mid-May and early June after 232-day gestation. A mother will have either one or two (and occasionally triplets), depending on range conditions. These calves will remain with the mother for 8 to 12 months (occasionally up to 24 months); however the mother eventually will chase them away in preparation for rearing the next calves.

Moose move around a limited area and have special places for wintering, calving and breeding. Moose eat all year. In summer they will fed on pond weeds, grasses sedges, forbs, and the leaves of aspen, willow and birch. In winter, they switch to willow, birch and aspen twigs.

Moose depend on their hearing and sense of smell to warn of danger; main enemies are wolves and bears. Calves are particularly susceptible to predation. A large, healthy adult moose can usually fend off an attack or escape so the very old and very young are the ones most vulnerable. Moose can be affected a parasite that can cause blindness, paralysis, disorientation and even death. Moose rarely exceed 16 years of age in the wild.

Northern survival

Moose are strong swimmers and can move quickly and quietly through the woods to escape from danger. Their heavy winter coats protect them from the cold. They will stay out of the wind by moving to wooded areas and will conserve energy by avoiding deep snows. This latter survival strategy can get them into trouble when they bed down on sled dog trails or on railroad tracks.

Tracks

Tracks are similar to deer but much larger; adults can be 4 to 7 inches long and 3 1/2 to 5 1/2 inches wide. Remember tracks will get larger in soft mud and snow. Strides can be 30 to 43 inches apart.

Other signs

Teeth marks on browse, rut pits (wallows), antler rubs on trees

Moose trivia

Moose are called elk in Europe.

Where to see:

Kenai National Wildlife Refuge, Denali Highway, Denali National Park, Chena River State Recreation Area, Campbell Tract (Anchorage, winter).

Muskox
© Alissa Crandall

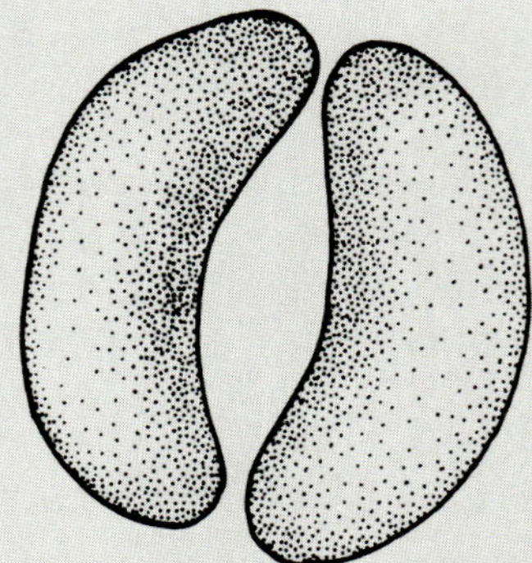

Where to see

The easiest place to see muskox in the wild is along the Dalton Highway north of the Brooks Range. However, this area is currently limited to travel by escorted bus tour. Visit the Alaska Zoo in Anchorage or the Muskox Farm in Palmer.

Muskox

Ovis moschatus

Range/Habitat

Muskox at one time ranged throughout most of northern Alaska but were hunted to extinction there by the late 1800s. They were later reintroduced back to Alaska from Greenland in the 1930s and now are estimated to number about 1,750 (1987) statewide. Muskox are currently found on the North Slope, Nunivak Island and the Seward Peninsula.

Physical description

Muskox are long-haired, stocky herbivores with short legs and a short tail; they have been described as "one of the heartiest hoofed animals in the world." Both males and females have horns but males' horns are heavier and appear to cross the entire forehead. Mature bulls stand five feet tall at the shoulder and weigh 600 to 800 pounds; cows stand about four feet tall and weigh 400 to 500 pounds.

Life cycle

Muskox mate in between August and October. A single calf weighing 20 to 30 pounds is born the following April, May or June.

Muskox are social animals, staying together in herds up to 75 animals except during the rut when one dominant male will keep a group of up to 15 females and subadults together as his own. Like sheep, the bulls will butt heads to establish dominance. One of the pair will eventually tire, get a headache or otherwise give up and turn away. Headaches are unlikely though as the muskox brain is protected by four inch thick horns backed by a three inch thick skull.

Muskox eat grasses, forbs, sedges and woody plants; occasionally they will add some flowering plants like saxifrage and mountain avens to their diet.

Muskox are known for their cooperative behavior in defending themselves from predators. Unlike caribou and other herbivores who will depend on speed for escape, the muskox stand together tightly. If the threat comes from one direction, they will often form a line; one bull may make a short charge ahead. If the threat comes from several directions, the herd will form a circle with their backs together, eyes and horns facing all directions. Young are pushed to the center. This worked very effectively against packs of wolves for thousands of years but proved to be muskoxs' undoing when confronted with whalers shooting with modern rifles. A single muskox will try to back up against a large rock or cliff when threatened.

Northern survival

Since the hoofs don't facilitate digging through heavy snow to obtain food in winter, muskox remain in places where the wind has blown the snow free. They can run quickly and easily over rocky terrain.

Muskox exist comfortably in some of the most brutal climate on earth due to their natural adaptation: the short stocky body (with short legs, ears and tail) naturally conserves heat and is covered with a fine double-layered coat of the longest hair of any animal in North America; hair strands can reach three feet long. This long, outer coat breaks the wind and a short, fine under hair is the best insulation known. Their hair also protects them from mosquitoes and flies in the summer; only the eyelids and ear tips are vulnerable to biting.

Alaska Natives painstakingly collect the fur (which falls off naturally in summer) and spin it into qiviut, one of the rarest fibers in the world. Qiviut is very lightweight yet is said to be better than wool in its insulating properties. Muskox are now being raised commercially for their fur.

Tracks

Cloven hoofs averaging 5 inches square on the front and 4 inches square on the back.

Muskox trivia

Muskox are related to sheep and goats. Native Alaskans speaking Inupiaq call the muskox "oomingmak," meaning "the animal with skin like a beard."

Short-tailed weasel
© Gary Lackie

Range/Habitat

Weasels are found in most of the state except offshore Bering Sea islands and the islands in southeast Alaska. They actually are quite numerous but most people don't think to look for them. Weasels prefer forests, brush and open country and often traverse along stream banks in search of food. Weasels are usually land based but have been observed swimming and climbing trees.

Physical description

Weasels' bodies are medium to dark brown with a white underside in summer, changing to entirely white in the winter. To tell Alaska species apart, look at the tip of the tail: the short-tailed weasel *(Mustela erminea)* tail is black while the least weasel *(Mustela rixosa)* tail has only a hint of black on it.

Weasels have long slender bodies. The short-tailed can be up to 15-17 inches long and will typically weigh seven or eight ounces. The least weasel is 10 inches long and weighs only three ounces, making it the smallest living carnivore.

Life cycle

Weasels are solitary hunters, living alone except during mating season. They breed in late summer but can have delayed implantation for six months; three to 10 babies are born in May or June, depending on latitude. Nests, often lined with mouse hair, are built in places such as small rodent burrows, rock outcrops or under buildings. The young will stay close to home while learning to hunt; they will go off on their own in the fall.

Weasel

Mustela sp.

A weasel's favorite food is mice but it will also eat small birds, fish and insects. The long thin body extends to the skull, allowing it to squirm down into the smallest of holes to extract its prey.

Weasels are considered fearless and very efficient predators; they can kill animals up to five times their size such as muskrats, waterfowl and squirrels. This is fortunate as weasels are also known for their high metabolic rate; this means an individual has to eat the equivalent of 1/3 or more of its body weight every day, leading one naturalist to call it "the most blood thirsty of all the mammals." This results in a tremendous expenditure of energy, so much so that they hunt by day and night and rarely have time to rest. A female feeding young will capture about four mice a day.

As you can see, food is very important to a weasel. Even when food is plentiful, it keeps on killing, storing the food for future use. It will not hesitate to confront larger animals and there are reports of attacks on humans that got between a weasel and its food cache.

Northern survival

The seasonal change in fur color is a double advantage; it helps a weasel hide from its prey but also keeps it from being hunted by its enemies such as horned owls or martens. If it has to run for its life, the black tail spot serves as a distraction that might allow it to escape. When all else fails, weasels can emit a pungent odor as a defense of last resort.

Tracks

Weasel paws are heavily furred. There are five toes on each foot but only four toes will print; overall size is less than one inch square. The hind feet tend to hit on the front tracks or just behind, indicating bounding gait.

Other sign

Food caches (for example, piles of dead mice), scat

Weasel trivia

The genus name *(Mustela)* means "one who carries off mice."

Where to see

Ermine avoid wetlands, preferring upland habitats like meadows, woodlands and mountains; its presence means there are a lot of rodents nearby to serve as a prey base. Weasels often will travel over a set route; watch for their tracks and sit patiently. You may also find them living under old cabins; such cabins definitely will not have a mouse problem.

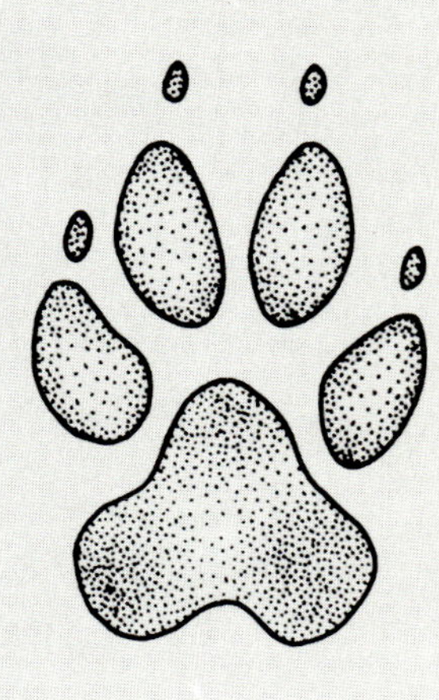

Wolf